KREW ENTERTAINMENT GROUP

PRESENTS

D.R.O.N.E.

KEVIN WHITAKER

McClure Publishing

MCCLURE PUBLISHING, INC.

COPYRIGHT © 2020

Kevin Whitaker for McClure Publishing, Inc.

All rights reserved. Printed and bound in the United States of America. According to the 1976 United States Copyright Act, no part of this book may be reproduced or utilized in any form or by any means, electronic or mechanical, including photocopying, recording, or by any information storage or retrieval system, except by a reviewer who may quote brief passages in a review to be printed in a magazine or newspaper, without permission in writing from the Publisher: Inquiries should be addressed to McClure Publishing, Inc., Permissions Department, 398 West Army Trail Road, #124 Bloomingdale, IL 60108. First Printing: March 30, 2020.

This book is a work of fiction. Any references to historical events, real people, or real places are used fictitiously. Names, characters, places, and events are products of the author's imagination. Any slights of people, places, belief systems or organizations are unintentional. Any resemblance to anyone living, dead or somewhere in between is truly coincidental.

ISBN-13: 978-1-7347595-0-1
LCCN: 2020905541

Cover Design by Kevin Whitaker

To order additional copies, please contact:
McClure Publishing, Inc.
https://www.mcclurepublishing.com
Books@McClurePublishing.com
800-659-4908

Krew Entertainment Group
http://www.KevinWhitakerBooks.net
krew40ent@gmail.com

First, I have to thank God for my life and the
abilities he has provided me.

Second, the family and friends God have given
me over many seasons.

Third, there are so many people I want to thank
individually, however, that would be a book in
and of itself.

Finally, with that being said, thank you all for
your love and support.

Know this if you read this until the end, I am
speaking to you.

Kevin Whitaker

INTRODUCTION

It is a warm fall day in Northeast Africa. A small tribe in Somalia is out harvesting their crops. The Sun is shining brightly when a sudden shade appears from above overshadowing the workers out in the field. The workers slowly turn to see the sudden shift of shade. The tribal leader already has his head up looking to the sky yells out, "CARAR (Run) UFO!!!" Instantly the tribe responds scattering in all directions quickly taking cover from this large Unidentified Flying Object.

Over in the United States of America, one of the largest gaming companies in the country is advertising their latest sensation, DRONES! The room is filled with executives from around the world to catch a glimpse of the latest technology in the gaming world. DRONE!

D.R.O.N.E.
<u>Death Rules Over National Enemies</u>

Klaevion Electronics has developed a new war game Drones! A contest is being held at McCormick Place®. The contestants have been selected through an online competition. This high-level war game has several people competing on one hundred sixty-five-inch monitors all around the room. After hours of competing, the contest has been narrowed down to the elite eight players of the game.

At one of the Drone stations Angel Chung, seventeen, very pretty with cocoa butter skin, tomboy style, has made the elite eight. No female has ever made it to the elite eight. Females always bottom out in thirtieth place when competing against male players.

She stands beside one of the board members with a big smile on her face convinced that she is going to win it all. Angel is matched up against Bryant Linton, Heavy Metal type, white kid with long hair and a leather jacket no matter the weather.

Bryant is contestant number seven. He is greatly confident that he could beat anyone. Bryant stands and just nods his head as the board member announces who he is matched up against. Angel pokes out her lips unfazed by Bryant's shenanigans. Bryant stares directly at Angel. Angel extremely focused, cannot wait for the opportunity to show what she is made of in the life of DRONE.

The C.E.O. of Klaevion Electronics Mr. Roosevelt Johns along with C.O.O. Mr. Kevin Alex, Jr. are in attendance as part of the three-judge panel. The third judge is from Tech Tech Co. a subsidiary of Heywire Electronics, Ms. Nicole Johnson.

Mr. Roosevelt Johns was the second child of his three siblings. Roosevelt had two sisters who were very protective of him growing up in the South Suburbs of Chicago. In high school, Roosevelt was a C average student with very minimal popularity. He was introduced to a stock trading class that would forever change his life.

Back in high school, Roosevelt's junior year, he had a math teacher named Mr. Creed. Mr. Creed was not only a math teacher he also taught stock trading classes on the weekends. Mr. Creed notices that Roosevelt was hanging around the class because normally he would be the first student out when the bell rang. Mr. Creed began gathering material he wanted to look at before heading to lunch and Roosevelt walks up to his desk.

"What's up Roosevelt? You're usually the first student out the door."

Roosevelt ponders, "Well...."

Mr. Creed encourages Roosevelt to speak, "Well?"

"I heard that you teach stock classes."

"Yes. I do teach stock classes on the weekends."

"I was wondering if you could teach me how to trade stock?"

Mr. Creed picks up his things, "Come and walk with me before you're late to class. Why do you want to know?"

"I was thinking about a career, and you got swag and all the honeys be on you."

Mr. Creed smirks, "It's not about the honeys. It's about the money."

Roosevelt smiles, "I'm feeling that."

The start of class bell RINGS!

"I tell you what Roosevelt. If you can pull that math grade up from a C minus into an A in the next quarter, we might make that happen."

Roosevelt excited, "I can get it done Mr. Creed."

Roosevelt walks in his next class like he just won the Jackpot Lottery.

Mr. Kevin Alex was the popular guy in school. He came from a family with money who happened to own property in the Loop area of Chicago. Kevin Alex the kid with two first names as they called him in high school was immensely popular. He wore the latest fashion and when he got his driver's permit, he got a new car. Kevin's parents always reward his success, so he was really a spoiled child.

Kevin's parents told him when he was a teenager that if he stayed out of trouble and kept his grades up that he could just about have anything he wanted.

Kevin recalls getting grounded one time by his Dad back in high school because he had stayed out after curfew. His parents told him it is good to be home before the streetlights come on.

D.R.O.N.E.
Death Rules Over National Enemies

One late afternoon, he was hanging out with his girlfriend Denise. They were at the park. He was pushing her in the swing which could be seen as lame by the guys today. Anyhow, he continued to push her and in a matter of moments the streetlights came on. He checked his watch. *It's only 5:00 pm*, he thought to himself. The streetlights never come on this early, so he kept hanging out with Denise.

Denise had gotten tired of the swing and decided to stop swinging dragging her feet through the mulch. Some mulch popped up and got into his eyes. Immediately, he stepped forward and Denise popped him with the swing. Kevin yelled, "OMG!" She quickly got off the swing to see to his well-being. Denise grabbed his face with both hands looking into his eyes. She could not see anything.

"You will be just fine." Denise exclaimed.

Then she kissed him and ran off yelling, "THE STREETLIGHTS ARE ON!"

Kevin took a minute trying to get himself together. By the time he got home, his Dad was sitting on the porch.

Mr. Alex Sr. angry walked in behind Kevin, "I thought we told you to come in before the streetlights came on."

"Have you even listened to what the boy has to say?" Mrs. Alex asked.

"Go to your room and don't come out until I tell you. You got a lot of nerves coming in here past curfew eyes all red."

Kevin tried to explain, "Mom, some mulch got in my eyes. I checked my watch and it was...."

Mrs. Alex holds back her laughter, "Baby that's the best you can come up with? Gone in your room we'll talk later."

Later that night Mrs. Alex goes into Kevin's room. "Baby you okay?"

"No mom. You guys wouldn't even listen to me."

"I know your Dad can be stubborn at times. You are grounded for a week, no computer or outside activities."

After a couple of days later, Mrs. Alex sat down to hear Kevin's story. She was convinced that Kevin was telling the truth. Unbeknownst to his Dad, Mrs. Alex had given Kevin back his computer until his Dad got home. She did not want Kevin's father to know.

Ms. Nicole Johnson was a hardworking student in college. She had nine siblings growing up in a house where they shared everything from clothing to food. It was Nicole's determination that led her out of poverty. Nicole went from working at Heywire Electronics to launching her subsidiary company Tech Tech Co. At the age twenty-nine, Ms. Nicole Johnson was on top of her game. She is incredibly attractive and always fashionable with Gucci, Prada, Louis Vuitton, and Versace.

Mr. Roosevelt Johns, Mr. Kevin Alex and Ms. Nicole Johnson all take a seat at the judges table as the battle between Angel and Bryant is about to begin.

This battle will be seen on large projection screens outside on the building also as part of advertisement. The panel of judges all chat briefly as the battle is about to ensue. One of the panelists, looks at the camera and says, "We will be right back after one of our sponsors commercial."

A Sponsor commercial appears on the screens, "JMFU, Just Mint 4 U, is a refreshing chewing sensation that freshens breath like no other leaving your mouth feeling clean."

Ms. Nicole giggles after seeing the commercial. "That's a pretty funny acronym – JMFU and clever marketing."

Mr. Alex sniggers, "That's not an acronym."

Ms. Nicole cuts her eyes at Mr. Alex, "Whatever JMFU!" She giggles looking at Mr. Alex.

Mr. Alex takes a quick breath check blowing into the palm of his hand. He checks to see if anyone is looking tossing a mint in his mouth.

Ms. Nicole smirks as he inserts the mint.

The battle begins and the first level is Border Patrol. Angel maneuvers quickly in and out accumulating points. Bryant is racking up points also with a series of combos of rapid fire and smart bombs as he wreaks havoc. Despite Bryant's arsenal display, he is still behind in points. They reach midway of the border and terrorists are seen trying to climb over the wall. Angel lets off a barrage of sidewinder bullets gaining points. Bryant swoops down with a smart bomb tallying up points killing everything in sight and blowing up tunnels around the border wall. His points skyrocket with just a couple hundred points ahead. Angel's swiftness has gotten her to the last level of combat with Bryant ahead on points. Bryant has to work against the clock now with Angel ahead by one level. He stops firing trying to navigate to the last level. As the intensity builds, Angel lays down massive attack when she realizes that Bryant will not be able to catch her then releases a smart bomb of her own racking up points. The subwoofer vibrates with an explosion of the smart bomb throughout the speakers in the room. Time runs out and Bryant loses.

The crowd erupts as Angel crosses the finish-line. Angel celebrates with her Drone with a double loop and the crowd heads move up and down while watching the movement of the Drone.

Angel with a great deal of excitement jumps up and yells, "YEAH!!!"

Mr. Alex walks over and raises Angel's arm, "You've done it young lady."

Bryant looks at Angel in a state of confusion as everyone congratulates her on the victory. Angel is enjoying all the attention shaking hands and taking pictures with guests.

Luke Lucas a/k/a Lucky Luke, age sixteen, black and very heavy into Hip Hop fashion, tall and extremely nice-looking kid is contestant number five. He looks at Angel and nods his head in approval as he listens to rap music through his headphones. Angel gives Lucky Luke the stare down. He chuckles.

Over at the United States Department of Defense (DOD), Admiral Greene was in attendance with five members of Homeland Security. Admiral Greene speaks with a real stern voice as he approaches Mr. Johns and Mr. Alex.

"Gentlemen! Where are we with our little project?" Asked Admiral Greene. Admiral Greene had a noticeably confident look that was quite intimidating standing at six feet five inches tall and two hundred fifty pounds. Mr. Johns pauses for a second looking up to the towering Admiral.

"Things are moving right as planned Admiral," exclaimed Mr. Johns.

Admiral Greene cracks a smile, "Yes, that's what I want to hear. This could prove vital to our national security." Admiral Greene glances over in Director Chris Mills' direction seeking his approval. Director Mills just slightly under six feet tall, mid-fifties, bald and slightly overweight rocks nervously.

"Yes, this project would work wonders and serve our country well." Director Mills stated.

Mr. Alex with the utmost confidence, "I assure you good gentlemen everything is going according to plan."

Admiral Greene looks down at his cellphone and types out a couple of reminders for himself.

"That's great news to hear that Klaevion Electronics and Homeland Security may have a deal in place"—he looks at Director Mills—"Mills, do you have any questions about anything?"

"No, not at this time Admiral. I'm sure somewhere down the line that there will be questions I want answered." Director Mills stated nonchalantly.

"Okay then. If there are no further questions this meeting is adjourned."

Admiral Greene slides back from the table. Everyone starts to pack up to disperse putting away computers and tablets. Admiral Greene walks over to Director Mills placing his hand on Director Mills' left shoulder.

"Mills, I need a word with you alone." Admiral Greene requested.

"Sure Admiral." Mills takes a seat allowing the room to clear out.

Admiral Greene leans forward standing over Director Mills, "This is surely a matter of National Security, is it not?"

Director Mills glances up at Admiral Greene, "Yes Admiral."

"You need to stay informed on every step taken until this project is complete. Are we clear?"

Director Mills stands looks Admiral Greene in the eyes, "Yes Admiral, crystal clear. I would not have it any other way."

Admiral Greene speaks with the upmost confidence, "We'll be the first to deploy such action using 5G. This real-time technology will give us the super-power that this country was built on."

Director Mills says, "Let me be blunt here. It has been reported that 5G causes cancer and mutation of human cells. In the bureau opinion, this technology will be a danger to all mankind."

Admiral Greene's forehead wrinkles, "Just as there is data saying it is dangerous there's data supporting 5G can be revolutionary. Know this real-time technology is the wave of the future and either we will be first or last. However, I do respect your position."

Director Mills takes a deep breath looking at Admiral Greene. "I do see where this could benefit Homeland with border patrol with real time Drones."

D.R.O.N.E.
Death Rules Over National Enemies

Admiral Greene with a devious grin extends his hand, "Glad to have you on board with this new technology."

Admiral Greene speaks in a low tone, "Mills … as far as I'm concern the only thing, I care about is speed. 5G operates at one hundred times the speed of 4G. We're talking real time."

Admiral Greene places his hand on Mills shoulder and squeeze just slightly, "Get it done!"

Director Mills looks up at Admiral Greene and nods in agreement.

* * *

Semi-finals day at the McCormick Place® and the room is packed with gaming companies and fans from all over the world. This Drones competition has taken on a life of its own.

Lucky Luke strolls through the crowd with his wireless KREW Bangers™ headphones listening to Hip Hop. Luke just takes it all in with coolness as he embraces the moment. Luke's swagger has him feeling invincible. He turns up the volume on his iPhone going into his own little vibe as the lyrics of the song speak to him, "Homie we don't play around … round … round. Most of my Hitters from the town … town … town, and we'll shut this trap down … down … down…, so you better move around … round … round…."

Luke approaches the stage where his opponent awaits his arrival. Ace the first-place winner in the last gaming competition. He is a twenty-year old white, arrogant, and fearless ex-football player All American Running Back whose career was cut short by an injury. Ace stares at Luke from the time they locked eyes until Luke reaches the stage.

Luke comes from a two-parent home. His Dad works for the city of Chicago as an electrical engineer, and his Mom is a registered nurse. He grew up in the downtown area called the loop. He got the name Lucky Luke, because his friends always told him that he was lucky for one that he lived downtown.

Luke as an only child got heavy into online gaming. This gave him an opportunity to meet other people. His Mom and Dad worked hard to provide a great life for him, but what he wanted was a family. It was at that point that Lucky Luke found an online family in the game of DRONE.

When Luke was younger, one hot summer day, him and his friends Eric, Brandon, and Shawn were at the Lakefront Skateboard Park. All the mothers agreed to drop them off at the park. Luke's mom, Michelle, stayed in the car to catch up on paperwork. She would always watch them in the park when they were young teenagers and warned them about the high board.

Brandon was the oldest by a few months, and he would always advise them as the leader of the bunch.

They all began skateboarding one at a time hitting the tallest ramp. Eric flew up the ramp at full speed doing a double twist at the top with a perfect landing. Shawn went right after Eric with a single twist falling at the bottom. The boys all start laughing.

"Dude, you alright?" Luke asked.

Shawn looked up at Luke, "Bro ... that was nothing."

"You guys are crazy. I'm not attempting any of those stunts." Brandon said.

Luke was more of the daredevil, "Brandon tripping. Watch this triple twist."

Brandon takes off on his skateboard circling the skating rink and returns to watch.

"Luke, your mom told us to stay off the high board." Brandon said.

"I got this...." Luke takes off at full speed maneuvering his skateboard up the ramp. He hits the end of the ramp reaching down to hold the board in midair."

Brandon, Shawn, and Eric all scream, "WHOA!"

Luke tries to stick the landing just as his skateboard touches the ramp. The front wheels pop off. Luke is tossed in the air. The skateboard shoots down the ramp at full speed with Luke sliding full speed behind it.

"NOOOO!!!!" Shawn, Eric, and Brandon screamed.

Eric takes off running going to get Luke's mom.

Brandon and Shawn stood over Luke.

"Luke isn't moving." Shawn said.

"Dude say something. Luke, can you hear me. Luke ... Luke." Brandon called out.

Shawn looks at Brandon, "Bro ... I think he's dead."

Everyone gathers around looking to see what is going on.

Luke shakes his head trying to regain his senses, looks up at Shawn, "You think I'm dead."

The crowd disperses and Michelle is standing over Luke. "You okay son?"

"Yeah mom, I guess that I forgot to tighten my wheels on the board."

D.R.O.N.E.
<u>Death Rules Over National Enemies</u>

"DIDN'T I TELL YOU BOYS TO STAY OFF THAT HIGH RAMP? GET UP AND GET YOUR THINGS BEFORE I LIGHT INTO YOUR TAIL. YOU MEAN TO TELL ME BRANDON WAS THE ONLY ONE WHO HAD ENOUGH SENSE NOT TO TRY IT. BRANDON GET YOUR THINGS. LET'S GO SHAWN. I'm calling all your parents when we get in the car. GET UP SON!"

Brandon walks with Michelle up in the front. Eric and Shawn walk with Luke.

"Man, if my mom had told me not to go on that ramp and I did, she wouldn't care that I was hurt ... she would come right over and punch me in the chest." Shawn said.

Eric whispers, "Yeah ... my mom would have gone nuts on me Bro. You lucky."

Shawn nods his head and says, "Yep Lucky."

Everyone is in the car while Michelle is driving. Luke sits quietly in the front seat.

"Son, I hope you learned your lesson. You guys are lucky no one really got hurt."

Brandon looks at Luke in the passenger rear view mirror and whispers, "Lucky!"

"After we get some ice cream, I'll drop you boys off at home." Michelle said.

Shawn leans forward tapping Luke's arm. "Lucky Luke."

* * *

Mr. Johns, along with Mr. Alex stand on the stage holding the largest check ever to be given out in gaming history. Mr. Johns approaches the microphone wearing an Armani® suit with Gucci® loafers. He speaks with a great deal of energy, "AS YOU MAY KNOW OR MAY NOT KNOW. THIS COMPETITION HAS TAKEN ON A LIFE OF ITS OWN. IT'S THE MOST RECOGNIZED GAME IN THE WORLD THANKS TO YOU GUYS, AND GIRLS. DRONES HAS EVERYONE ON SOCIAL MEDIA GOING CRAZY. CAN YOU DIG IT!!!"

The crowd ignites!

Mr. Alex steps up to the microphone holding the other poster size check for one million dollars. He speaks loudly, "ON BEHALF OF KLAEVION ELECTRONICS THESE CHECKS WILL BE PRESENTED!!!"

At that moment two beautiful women come out to raise the checks. Lucky Luke and Ace both admire the girls with smirks on their faces.

Mr. Alex continues, "Fourth place will receive a check in the amount of twenty-five thousand dollars!"

D.R.O.N.E.
<u>Death Rules Over National Enemies</u>

The crowd erupts with cheers as the two girls walk around with the poster size check of twenty-five thousand dollars.

Mr. Johns sarcastically speaks, "Alex, no one wants to hear about fourth, third, or second place winnings. They want to hear about first place!"

Mr. Alex, "So you think Mr. Johns? Well in that case......FIRST PLACE IS ONE MILLION DOLLARS!!"

The room explodes with screams as the girls walk around with the oversized check of one million dollars. The room is filled with energy and anticipation for the next round.

Mr. Alex tries to speak over the crowd, but they are *turnt* up! "Let me have your attention so we might be able to see who that one million-dollar winner may be? To the right we have Lucky Luke, and on my left, we have Ace. 'Ace is there something you would like to say to Lucky Luke?'" Mr. Alex turns the mic in Ace's direction. The crowd is silent as Ace is about to speak.

Ace with a slight chuckle, "Yeah, after my injury I thought that I would never have a chance to make a million dollars. The entire time of my recovery I was thinking of ways to make it happen. I'll be a goofy if I let some kid name Lucky Luke take that away"—looking directly at Luke—

"after this game they'll be calling you Luke, because your luck just ran out!"

The audience laughs amongst themselves.

Mr. Alex chuckles, "Wow! This battle just intensified. Lucky Luke is there something you want to say?"

Lucky Luke removes his headphones. "Yeah, Ace is nobody in this space … if you think about it, he's on pace to get his ACE whooped in this competition!"

The crowd goes wild. "Luke just came for you Ace." Mr. Alex stated.

Ace nods in agreement.

Mr. Alex surprise, "Okay!" He laughs, "This competition is heating up. Okay, before we go any further our sponsors are KEG (Krew Entertainment Group), KREW Bangers™ headphones, BOA BunS of America, Winners Ice—make sure you get that Ice, and Just Mint 4 U (JMFU) for that daily fresh breath."

Ace has a smirk on his face as Lucky Luke walks over to the Drone stations.

The media is largely in attendance. The number one news station W.E.R.K. has center stage. Standby for the countdown before they broadcast live as the camera man counts down. He points his finger signaling ACTION!

"Hi, I am Larita Duncan reporting live from McCormick Place®, downtown, Chicago. It's a beautiful day in the city and inside this room is full to capacity. What would have people indoors on such a beautiful day? How about the largest gaming competition in history DRONES with the first-place prize of one million dollars cash. I'm standing here with one of the executives of Drones, Mr. Johns of Klaevion Electronics.

"Mr. Johns how were you able to assemble such a high volume turn out?" She turns the mic towards Mr. Johns for his comments.

Mr. Johns with a million-dollar smile speaks, "As you know electronic gaming has gone global and our mission is to link everyone worldwide with Drones. Our online process has made it possible for anyone with internet access to compete and the outcome has been lucrative. Therefore, in our success, we designed a system that gives back to our customer base."

"Wow! That's exciting. One more question Mr. Johns. 'Why didn't you spread the million-dollar prize out over the top four contestants, instead of giving a million-dollar prize to one individual?'" Ms. Duncan asked.

"Our model here at Klaevion is winner takes all. The twenty-five thousand-dollar check is for runner up and travel cash."

Ms. Larita Duncan nods in agreement, "Thank you, Mr. Johns for your time."

Mr. Johns turns back to the stage.

"Woo this crowd is electric. I can barely hear myself speak over the music and fans. I'm Larita Duncan reporting live from W.E.R.K. T.V. Back to you Dan in the studio."

Dan Smith is the anchorman at studio W.E.R.K. "Thanks, Larita. We have breaking news. This story just in from our affiliate station KREW T.V. (footage is being shown). There was some type of attack in East Africa today. No one has taken responsibility for this attack. However, you can see villages burning and tribes scrambling to gather any of their possessions. This is just disheartening. I want you to know whoever is behind this, the United Nations will not tolerate this type of behavior. I'm Dan Smith. Now a word from our sponsors."

Sponsor—soft background music starts playing behind a bluish black screen when a deep voice starts speaking while words go across the screen—"looking for excitement look no more KEG (Krew Entertainment Group®) the epic center of entertainment."

Back to the competition Lucky Luke has just crossed the finish line defeating Ace. Ace stands there in disbelief mugging Luke. Lucky Luke is standing with Mr. Alex with his arm raised. Lucky Luke nods his head agreeing with the loud crowd of people. They chant, "Luke … Luke … Luke…."

Lucky Luke raises his fist in the air with a big grin looking in Ace's direction. "Yeah, I'm in the building!"

Director Mills stands in the back of the room watching on the monitor. He has seen the entire competition. He pulls out his cellphone calling Admiral Greene.

"Hello!" Admiral Greene answers.

"Admiral, this is Director Mills. The competition is not a competition. This kid Lucky Luke easily beat one of the top opponents." Mills said as he watched the monitors.

Admiral Greene sitting in his office starts to rock back and forth in his chair. "Thanks for the intel, Director."

With all the hype about the Drones competition and the ensuing battle Lucky Luke appears on top Electronics Magazines. Ace was Luke's opponent and he did not like all the hype surrounding Lucky Luke. Ace was featured on the front cover of Amazing Electronics as one of the top opponents in the competition.

The competition is heating up on social media. Banners about the competition going across several pages while numerous people are sharing it on their pages. Angel does not understand why all the talk was about Luke and Ace when she defeated a top contender.

Angel heads home with her victory on her mind. She sits on the edge of her bed playing Drones. Her backpack is on the bed behind her.

Mrs. Chung enters the home carrying packages, sits them on the counter, and grabs the mail. She quickly shuffles through the mail,

"Bills ... bills ... bills!"

She pauses and listens closely. She nods in disappointment walking towards Angel's room.

"You better not be playing that game!"

Angel mimicking her Mom "You better not be playing ... oops!"

Mrs. Chung opens the door and catches Angel. She stands over Angel tapping her foot on the floor. Angel quickly drops the controller and grabs her book from her backpack.

"Mom! You suppose to knock first."

Mrs. Chung with an Asian accent, "You don't tell me young lady, I tell you. I will knock as soon as you get your own place. Turn off that television now!"

Angel opens her tablet looking at her mom.

"That's just so disrespectful busting up in here like the police."

Mrs. Chung winks at Angel, "I am the police in here. I love you though." She turns to exit the room closing the door behind her.

Angel yells from her room, "That's why your skirt is too tight."

Mrs. Chung quickly opens the door back. "No Sweetie my skirt isn't too tight, but those jeans and T-shirt you wear are unappealing or *racket* as you kids say today," Mrs. Chung said with a slight giggle.

"Besides that, you are not a little girl anymore. You need to start dressing like a young lady."

"No ... no ... and no... you want to be trying to act all hip. Mom please never talk like that around my friends you'll embarrass me. *Racket,* really mom?"

Angel falls back on her bed kicking her feet up in the air.

Mrs. Chung chuckles, "Now that's embarrassing acting like a two-year old."

She turns to exit the room once again. Angel immediately grabs the controller getting back to her game of Drones.

* * *

Mr. Johns is sitting in his office talking on the phone. His secretary enters the room. He signals for her to give him a minute.

"Thanks a lot. We'll talk later" Mr. Johns ends the call.

"Okay, Ms. Cole what can I do for you?"

"Mr. Alex has arrived. Shall I send him in?" Ms. Cole asked.

Mr. Johns replies, "That will be fine, send him back."

Ms. Cole turns to exit and pauses, "Will you need anything else, Sir?"

"No, just send Mr. Alex on back."

She walks out of the office straight to her desk. "Excuse me Mr. Alex."

Mr. Alex texting someone while answering, "Yes, Ma'am."

"Mr. Johns is ready to see you"—Ms. Cole states while waving her hand in the direction to lead Mr. Alex—"you can just go on back."

"Thanks!" He stands and walks to the office of Mr. Johns.

Mr. Alex enters the office and Mr. Johns stands to greet him with a firm handshake.

"How are things coming along?" Mr. Johns asked.

Mr. Alex spoke with a little concern, "I have to say things are just fine."

"Have a seat." Mr. Johns stated.

Mr. Alex takes a seat next to the desk near Mr. Johns .

Mr. Johns with a big grin on his face.

"This global competition was genius. The stock of this company just keeps on rising with all the promotion around Drones." Mr. Alex exclaimed.

Mr. Johns grinds his teeth not knowing where the conversation was headed.

"Yes, I would agree. However, I'm a little more concerned about the Department of Defense contract. Where are we on that?" Mr. Johns asked.

"Chris Mills stopped by my office yesterday. He told me that he attended the competition and liked what he had seen. He and I spoke briefly then about the contract." Alex's eyes widen as Mr. Johns sat straight up.

"I do believe that Mr. Mills was quite impressed with our online registration for Drones. We have over a million gamers online at five dollars each, every month to stay in the competition. This global market has set new boundaries for our company."

Mr. Johns flips open his laptop, "According to my numbers we are more like ten million sold with repeat customers trying to stay in the competition.

"Since the last competition went viral our numbers are more like twenty million. Everybody wants a part of the Drone experience."

Mr. Johns smiles as he rubs the palms of his hands together.

"That's excellent news Mr. Alex."

At that very moment Mr. Alex's phone vibrates. He glances down at his cellphone. The message reads, THIS IS YOUR SECOND INSTALLMENT ON THE DOD CONTRACT $50 million. Mr. Alex's eyes light up as he reads the message then an instant look of fear.

"They just wired the money to our overseas account."

Mr. Johns erupts in a joyful laugh.

Mr. Alex afraid, "This is enormous. What happens if we can't produce?"

"We face a firing squad," Mr. Johns said jokingly.

"We are already succeeding." Mr. Johns said happily.

Mr. Alex reflects on Admiral Greene's last words *failure is not an option.*

"The marketing strategy to let gamers represent their own country and then allow them to travel to the United States is genius." Mr. Johns replied.

"No failure here."

*　　*　　*

At the hottest Hip Hop restaurant in town the gamers and skateboarders have one thing in common which is the love of Hip-Hop music. The vibe is awesome with video monitors showing music videos and this restaurant official game show R' U' Really Hip-Hop. The place is packed with older teens and young adults. Angel and Lucky Luke were seated at the same table for the show. Angel's table of four orders a pizza and a pitcher of soda. The online show is about to begin.

Lucky Luke is watching the music videos until the food arrives. Kyla, a friend of Angel's, sat at the table with them. Jamarcus was a prize winner for the game show. He is the incredibly quiet type and shy despite being handsome.

Angel takes a bite of her pizza gazing at the stage and lights. Lucky Luke stares at Angel thinking about *how beautiful she looks* to him. Angel turns to grab her soda catches Luke staring at her. She instantly gets defensive. "WHAT?"

Lucky Luke quickly looks away, "What?"

"Do I have some food on my face or something? I mean the way you're staring at me. That's irritating."

Lucky Luke nervously thinks to himself, *Dang she caught me.* "No, there's nothing wrong with your face."

Angel gives Lucky Luke a hard stare with a frown. The light on their table begins flashing indicating that your table has been selected to participate in the game show. Angel's facial expression quickly converts to excitement.

"We've been chosen." Kyla stated.

"That's what's up." Jamarcus added.

"Woo hoo!" Angel shouted.

Lucky Luke just sat there with a nonchalant attitude as the Host takes center stage. The Host is a tall, black guy, who happens to be a comedian. He is sharply dressed and well groomed. "How's everybody feeling this afternoon?"

"Good!" The crowd answered in unison.

"My name is Gemini, and I'll be your host for the show. I want to know (the audience joins in) R'U' Really Hip Hop?"

"I'm the king of fun ... and on that note, I see some funny looking people up in here. Just remember nothing I say up here is personal because I don't know you. However, yawl some funny looking people like this guy right here." He points at himself.

The crowd laughs.

"I have a gift for everybody in here ... because I don't know who the most lit person in the room is, I put them on standby."

The audience all look around the room at one another.

Gemini walks to the end of the stage. "Table seven over here looks like their *turnt* up."

Table seven yells loudly.

"Yep ... table seven it is. I'm going to need you to go to your favorite store and pick something out you want and call me. Once I hear from you then that gift is yours. At that very moment when I hear from you that gift will no longer be on standby.

"Wait a minute. Now take a selfie next to it with a big smile like this"—Gemini smiles and takes a picture with the audience in the background—"hold on hold on … listen up. Now that gift is no longer on standby because you got a picture standing by it. Buy it yourself.… Get yo wanting something for free self-up out of here."

The customers burst out in laughter.

Gemini walks the stage trying to hold in his laughter, "Okay now who wants to be a millionaire? Wrong show seriously.… Now, who is ready to play the game. In unison the customers and Gemini ask, "R'U Hip Hop?

"Now most of us know the rules. But, for those of you who don't, when your table lights up, you'll have ten, twenty, thirty seconds to answer. The tablet will begin the countdown as everyone else watches on the T.V. monitors.

"Kathy is our contestant and she needs your help. Whatever points you assist Kathy with, your table will get those same points which ends in MONEY! Okay table five you have thirty seconds after Kathy reads you the question, and remember no cellphones. Phones get you banned around here."

D.R.O.N.E.
Death Rules Over National Enemies

Gemini passes the mic to Kathy. "Okay, Table Five, I need to know what year did T.I. release his first C.D. not mixtape?"

The clock ticks quickly and loudly. Angel has no clue. Jamarcus ponders on it. Luke sits back calmly. Angel screams at Luke, "Do you know the answer?" The clock ticks less than five seconds.

"I think it was three years after the turn of the millennium 2003." Luke answered.

Angel screams at the top of her lungs while selecting answer D on the tablet, "2003 … 2003!!"

Gemini frowns then shakes his head indicating no. At that split-second he yells out slowly, "I feel so bad for you—then quickly says— "you're correct!" T.I.'s video plays for five seconds before going to the next question. Kathy is all excited jumping around on the stage.

Angel leans over the table giving Lucky Luke a peck on the cheek.

Jamarcus smiles, "Woe! I should have had that answer."

"What was that for?" Luke asked.

Angel replies, "Getting the right answer."

Luke smirks, "If that's all it took somebody should have asked me something sooner."

They all start laughing at the table. Angel playfully punches Luke in the arm.

The host Gemini walks across the stage with a big grin on his face "Okay that table was successful in helping you. Before you can move on to the next level you have to answer the next two-part questions correctly.

Gemini hands Kathy the mic again and she asks, "In what year did Young Chop appear on the scene of Hip Hop?" Second part of the question, "Name two of the three songs that helped put Chief Keef and Young Chop on the map?" (Hip Hop instrumental track plays).

Jamarcus says, "I got this."

"No, I know this. I sang Chief Keef's songs by heart." Kayla exclaimed.

The host Gemini starts smirking and says, "Well, well, well, you got this Jamarcus, and since Kayla is noticeably confident, we will have her answer one of the questions you cannot answer.

"Young Chop appeared on the scene in 2012, and one of the songs I remember he helped Chief Keef is 'I Don't Like.'" Jamarcus answers.

Kayla interjects immediately afterwards, "There are two more, but I'll give you one song. It is 'Love Sosa.'"

Gemini looks surprised with the knowledge millenniums know about Hip Hop and states slowly, "Your answers are correct."

Bells and whistles filled the room while Jamarcus and Kyla exchange numbers.

As everyone starts to exit the restaurant for the next taping of the show, R'U' Hip Hop. Angel just smiles as they walk toward the exit.

They step outside; the sun is brightly shining. Angel covers her eyes from the sun's glare as they walk pass the others who are trying to get in for the next taping of the game show. The crowd shoves one another trying to get out of the venue. Everyone is chatting about the show.

Ace is standing in line with a few of his friends waiting to get inside. Ace gives Lucky Luke a hard stare. Luke does not even acknowledge his presence. He just keeps walking. Angel giggles. Kyla explains to Jamarcus who Ace is. Luke drops his skateboard and scurry off.

Ace frowns throwing shade at Lucky Luke, "He's a Goofy!"

Angel stares at Ace with great disgust, rolls her eyes, and keeps it moving.

* * *

The next day at the Department of Defense, Admiral Greene is sitting in his office talking with Homeland Security, Director Mills.

"Mills, I think that it's time that we activate Phase III."

Director Mills confused, "Phase III?"

Admiral Greene with a devilish tone in his voice, "Yes, I have a couple young cadets for the Drone competition. These guys should easily beat whoever is in front of them. Besides, it won't hurt if they win a little money. So, I need for you to get them in the competition."

"Excuse me Admiral, you are aware it's the semi-finals, that task may be impossible."

Admiral Greene slightly agitated, "That's what we do. We make the impossible possible. I'm going to need for you to get it done. I don't give a damn how you make it happen, just make it happen. Lieutenant Belmer will give you the names on your way out. You're dismissed Director."

Director Mills unsure in his mind on how to get it done frowns and covers his mouth with his fist. He stands and salutes Admiral Greene turns to exit. He stops out at the desk to speak with Lieutenant Belmer.

"You have a package for me Lieutenant Belmer?"

"Yes Sir, I do have a package for you." She turns around in her chair removing the package from her file cabinet. "Here you go, Sir."

"You're very polite and beautiful." Director Mills said.

"Well thank you, Director! Is there anything else I could get you?"

All kind thoughts crossed his mind. "NOPE! That will be all"—he exits walking towards the elevator. On the elevator he opens the package reading out the names—"Henry Lee a/k/a Sho-Gun, Air force pilot, and Benji Smith a/k/a Black Rain, Naval Pilot"—Director Mills nods his head while stepping off the elevator speaking out loud—"they sure better win."

Benji Smith a/k/a Black Rain is a young man out of poverty-stricken Detroit. He was a troublemaker in the streets and after one arrest where his grandfather was involved, he made him join the Navy through a waiver. He became a naval pilot. He rose up in the ranking quickly because of a mental fortitude of not caring and a treacherous way of thinking which made him a threat to anyone caught in his path.

Black Rain spent time playing games online and became known for his online gaming presence.

Benji as a teenager got into all types of criminal behaviors from stealing cellphones, cars, and some minor drug charges. On his 20th birthday, he had to appear in court. His Grandfather Benji III was in court with him. Benji was sitting in the front of the courtroom awaiting his time to see the Judge. As Benji sat there he had a chance to witness other people being sentenced. The judge seems to be a no-nonsense type late 50s and white. Benji just sat there and listened thinking about himself and how this could play out for him.

The Court Reporter called out the case, "Benji Smith please step up to the front." Benji and his Grandfather rise and walk to the Judge.

Honorable Judge Carl Thomas instructs them, "Young man step to the bench. Dad stand to the right of Benji."

"I'm his Grandfather, Sir. He lost his dad at an early age."

"I'm looking over your criminal history here and you have three or four minor offenses. Young man turn around and look behind you."

Benji looks over his right shoulder.

"I asked you to turn around and slowly look behind you."

Benji turns slowly in a circle until he was facing the Judge again.

"What did you see when you turned around?"

"I see a lot of people and my granddad."

"That's not what I see. I see your future. Now turn again and look."

Benji gets nervous turning this time even slower.

The entire courtroom is staring at Benji the second he turned.

"Young man … I won't ask you what you saw this time. I'll tell you what I see as you turn around one more time slowly."

Benji starts to turn and Judge Thomas speaks as he turns, "I see a courtroom filled with police and sheriffs. I see people in handcuffs, I see people with Department of Correction jumpsuits or D.O.C. and people with little support or family members here to support them."

Benji completes the turn and is now facing the Judge again.

"I see your future if you don't change your ways. I was ready to sentence you today. On behalf of your Grandfather and the way he presented himself in the courtroom, I see him as a man. His attire is that of a man well dressed and well groomed. You young guys come in here with sagging pants underwear showing and think it's fashionable. Not in my Courtroom or any other profession that I know.

"I'm going to allow you to leave here today under supervision for twelve months. However, if you get caught doing anything no matter how small it maybe, you shall go to prison for the remainder of that time. Do you understand what I'm saying to you?"

Benji swallows his pride, "Yes Sir."

"Oh! Benji a/k/a Black Rain ... we do know who you are and where you are"—then pauses for a second—"this goes for everyone in this courtroom. People are not bad. Somewhere down the line they made a bad decision. Think before you act out any bad intentions. Benji, you may see the clerk. I never want to see you in my court again."

"Does the state agree with my ruling?" Judge Thomas asked.

"The State concurs with the Judge's ruling." States Attorney replied.

Benji takes a hard look around the courtroom and walks away with his Grandfather.

Benji said nothing the entire ride home.

"You're going to the Navy." Grandfather Benji said.

Benji did not contest the thought of going to the Navy, "Okay Pawpaw."

The other new contestant is Henry Lee a/k/a Sho-Gun an Asian kid with a great background whose grandfather and father had served in the Air Force as pilots. Sho-Gun just wanted to follow in his father and grandfather's footsteps and became an Air Force pilot. He earned his name from a karate competition that he had won after being injured. It was almost like a clip out of the Karate Kid.

Sho-Gun as clever as a fox and more dangerous than the world deadliest snake, became known as an online gamer executionist defeating any and everyone in his path.

At McCormick Place® the semi-finals continue on spring-break week. The place packed once again with gamers, media, and technology personnel. The competition can be seen live on the internet by subscribers. Lucky Luke is chilling over by one of the Drone stations watching others play until his battle.

Angel is enjoying the festivities as she meets and greets people.

The media station W.E.R.K. is filming. "Good Afternoon! I'm Larita Duncan reporting live at the McCormick Place® at the Drones competition. Never have I seen such a showing for a video game. This place is earsplitting which is drowning out my words."

The sound of Drones flying, shooting, and music mixed in with loud voices.

"Here to my left is one of the semi-finalists Angel Chung."

Angel steps in and waves at the camera, "Hi!"

Larita extends the mic and asks, "Angel, are you ready for the next round competition?"

Angel with a big smile, "Yes," Angel turns and walks away.

"Such a beautiful young lady. She is not intimidated by these boys or the game. She is the only female left in this competition. Go Angel! I'm pulling for you. We'll keep you updated as this competition gets underway. Back to our studio with Dan Smith."

"Good Afternoon, I'm Dan Smith and this is my co-host. The beautiful Latina, Maria Santiago."

"Elsewhere in Northeast Africa there had been a couple attacks in Somalia over the past week. No nation has taken responsibility for those acts of war; however, there has been many casualties reported." Maria Santiago states in her Spanish accent voice.

Meanwhile over at the competition the Drones showdown is about to begin. The owner of Klaevion Electronics, Mr. Johns stands center stage with Mr. Alex and Director Mills. Mr. Johns steps forward grabbing the mic.

"Hello everybody! It gives me great joy to see the house packed once again for Drones. I'm going to need your undivided attention for a moment for this announcement." Mr. Johns runs the palm of his hand under his chin.

The crowd goes silent when they see the expression of Mr. Johns on the monitors.

"First, I would like to thank all of you for making this competition a success. However, it is my job to make sure that we have the best of the best in this competition."

The room erupts in cheers.

Mr. Johns continues to speak, "It gives me great pleasure to bring you the Vice President of Operations and organizer of the Drone competition. Mr. Kevin Alex."

The crowd goes wild whistling, shouting, and clapping.

Mr. Alex smiles from ear to ear trying to speak over the audience. "Okay. In all my years of gaming, I have never seen such an electrified crowd. (Crowd Cheers!) I know you're ready for the semi-finals. ARE YOU READY?"

"YES!!!!" The crowd responded.

"As Vice President of Operations, it's my job to inform you of the changes that have been made here today."

The crowd strongly disagrees with hostile words.

Some yell out of the crowd, "What kind of change are you talking about? Making America great again?"

Director Mills abruptly grabs the mic from Mr. Alex.

"I assure you this will only take a few seconds of your time before this competition gets on its way. Two new players have been added to the competition."

The crowd erupts with profanity, "This BOGUS! What the hell is going on? Boo … Boo. Where they do that at? Change the rules at the finals."

In unison the crowd, "boo … boos!!!"

Ace yells out, "How can someone get in this late. You're bogus Dude!"

Larita Duncan asks her cameraman to start taping at that moment as the crowd starts getting irritated.

"I assure you these two guys have been tried and tested. They are two of the internet sensations. Because of their active military duties, they were not allowed to participate in the online competition. I know if I was playing the game, I'd want to play the best out there so there's no room for doubt. If one of you beat either one of these players you will win an additional $50,000, how about that for some extra excitement?"

The crowd quickly glances around the room at one another then start cheering loudly!

Larita Duncan taping in the background, "Audiences at home this has the look of anarchy."

Ace leads the crowd, "BRING'EM OUT ... BRING'EM OUT ... BRING'EM OUT!!!"

"YEAH...YEAH!!!!" The people responded.

"Let me introduce SHO-GUNNN!!!!" Mr. Mills shouted.

Henry Lee walks out wearing jeans and black T-shirt looking at the crowd over his dark shades. He nods his head yes to the crowd as he walks around the stage arrogantly.

"Next up is none other than Benji Smith a/k/a BLACK RAIN!!!!! Black Rain walks around wearing hip hop fashion carrying a backpack. Black Rain walks the stage fearlessly staring at his opponent.

"Angel will have the luxury to take on Sho-Gun, Henry Lee, and the one who seems to be the people's champ Lucky Luke will take on none other than the Grim Reaper himself Black Rain," Mr. Alex states.

Angel is furious staring up at Mr. Mills. Lucky Luke is unfazed by that task until he looks and sees how upset Angel appears.

A spectator yells out "What type of nonsense is this? Those Bozos should start at the back of the line like everyone else."

Chants of... "Angel ... Angel ... Angel!!" fills the room then the entire crowd joins in with chants of Angel.

Angel's look of disbelief turns into one of confidence waving at the crowd.

Mr. Mills continues, "Gamers, the moment that you have been waiting on quickly approaches us. Who will advance to the next round will it be Angel or Sho-Gun, the destroyer?"

Ace standing in the crowd with his friends recognizes Sho-Gun from social media.

Kelli, Ace's girlfriend whispers to Ace, "Is he any good, does Angel have a chance at defeating him?"

Ace nods his head, "It's going to be tough for both of them to win. Angel's no push over. Black Rain is even more dangerous than Sho-Gun. I played against Sho-Gun online in Operation Storm and for some odd reason he had all the killer combos to destroy me. I know now how he was able to get those codes."

Kelli with a look of concern, "How did he get the codes?"

Ace disgusted says, "The gaming company."

Tim, a good friend of Ace, was listening to the conversation, "DANG! It's about to get real ugly in here."

Ace frowns, "For sure."

Kelli smiles, "Ace you could defeat anyone of them."

Ace smirks, "If that was the case I'd still be in the competition."

Black Rain walks over and stands next to Lucky Luke. Luke with his Beats by KREW Bangers™ headphones on turns and just nods his head to the beat of the music while his eyes are slightly closed. Lucky Luke tilts his head in Black Rain's direction opening one eye to see Black Rain.

Mr. Mills speaks sternly, "Okay, let's establish some order to get this underway. I get it. Some of you are a little upset of this sudden change; however, it is what it is. This weekend will reveal who the number one gamer is in the world of DRONE and a first-place check for one million dollars, plus a $50,000 bonus."

The crowd is split as some of them chant Angel and the others chant Luke. Black Rain and Sho-Gun acknowledge one another with eye contact and slight head nods.

Larita is standing backstage when her cellphone buzzes. "Hey Dan, what's up?"

"The social media world has gone crazy about some last-minute change in the competition. What's happening?"

"Something about two last minute entries."

"Well, a lot of people are saying on social media that the contest is rigged. What do you think?" Dan asked.

"I've never seen anything like this and definitely for that type of money. Who knows? Got to go to the competition. The game is about to begin."

Center stage Lucky Luke sits in the ring and Black Rain a short distance away is at the next Drone station. Angel sits sideline observing when her and Luke make eye contact. Luke winks his eye at Angel, and she smiles. Luke makes a couple adjustments to his headphones and selects his playlist on his phone. Lucky Luke looks up; Angel winks at him.

Mr. Alex grabs the mic giving his best Michael Buffer voice, "GAMERS ... look alive ... look alive it's time to roar." The crowd rumbled like the sound of a stampede of horses.

Black Rain looks over trying to intimidate Lucky Luke. Luke smirks and nods his head up and down.

Mr. Alex asks, "Lucky Luke are you ready? Black Rain are you ready?"

The energy in the room is full throttle and the audience pumped.

"With all the viewers around the world and over one million online and the capacity crowd in attendance, let's get it cracking!" The crowd erupts!

Mr. Johns walks over whispers in Mr. Alex's ear, "What the heck was that get it cracking stuff?"

Mr. Alex chuckles and whispers back, "Just trying to be hip. 'Gamers, select your Drone.'"

Lucky Luke has selected Heron/Heron TP which is known for its sensors that can easily detect a threat. Black Rain selects Harfang Drone which is known for autopilot and sleek moves. If somehow the controller loses control at top notch speed, he can still fire his weapons.

"Gamers, get ready for the Death Course that no man has completed in the tournament."

The crowd goes silent when the track is selected.

"The person with the most points and distance will advance to the finals. Let the countdown begin ten ... nine ... eight ... seven ... six ... five ... four ... three ... two ... one!"

The moment everyone has been waiting on has begun. Everybody watches the screens closely as Lucky Luke takes a slight edge. The sound effects from the Drones fill the room as Black Rain lets off an arsenal gathering points.

Back over at W.E.R.K. Studios, Dan Smith sitting in the editing room staring at the clock and looking at some footage from some of these past attacks on the third world countries. Something has captured his attention with the Drone competition. He falls back in his chair folding his hands together on top of his head in deep thought.

D.R.O.N.E.
Death Rules Over National Enemies

Meanwhile, over at the Department of Defense Office in Chicago, Admiral Greene is in the conference room with members of the CIA and FBI.

Admiral Greene speaks, "Let me address the agencies concerns. We all know the importance of securing the Gulf of Aden and the Red Sea. Phase III has already been implemented to assure that our national security concerns are taken care of in this matter."

"Yes, on behalf of the CIA, we understand that this course of action that you have chosen will eliminate the threat. However, we will stand down and let your Department handle this problem. Nor will we assist you in this matter. This course of action you have chosen hasn't been approved. Therefore, the CIA will stand down. We will not assist you nor will we stop you from these actions."

FBI member speaks, "I concur with everything the CIA has stated. If any questions arise about this mission, the FBI will deny all knowledge about this action. I hope that we are clear on this matter Admiral Greene."

Admiral Greene very cocky, "I clearly understand your position, but there is Intel that suggests that an attack is being plotted as we speak on the American soil."

CIA member with a stern look, "We understand that threat to be true; however, the mission hasn't been sanctioned by the president."

"All precaution has been taken to prevent unwanted casualties. This means operation Clear Water Phase III is underway gentlemen." They stand in unison each Department shook hands in agreement before exiting the room.

Admiral Greene sits back down at the table in deep thought speaks out loud, "Plausible deniability, lames!"—shaking his head in disbelief—"everyone wants the reward, but none are willing to make the sacrifice."

Back over at the Drone competition. Black Rain has taken the lead over Lucky Luke. The audience stares at the screen as Black Rain starts to pull away from Lucky Luke. Black Rain with a devilish grin on his face senses that Lucky Luke does not stand a chance. Luke desperately trying to gather points opens up with an array of bombs and bullets to raise his strike points.

On the screen a large cargo ship appears under attack by pirates which Black Rain and Lucky Luke have to protect. Black Rain swoops down with Lucky Luke right behind him. Lucky Luke has more strike points but is behind on the course. Black Rain stops firing to pick up speed headed for the end of the course.

D.R.O.N.E.
<u>Death Rules Over National Enemies</u>

The sound of Drones at top speed is electrifying to the audience as they watch and listen. Headed for the finish line at top speed Lucky Luke is gaining on Black Rain quickly. Black Rain drops down in altitude picking up speed. Lucky Luke drops down behind Black Rain and is gaining on Black Rain.

The crowd sits ever so intensely waiting for the outcome. The low altitude buzzer and flashers are engaged flashing and speaking **(*pull up now you are traveling too fast at this altitude pull up now!!*)** Black Rain ignores the warning sensors. The crowd is in utter amazement never seeing anyone operate at this altitude most gamers have crashed.

Black Rain has Lucky Luke by a narrow margin nearing the end. Lucky Luke with a five-button combo gives his Drone a thrust pulling alongside of Black Rain. Black Rain smirks, putting together a four-button combo thrusting him back in the lead. They reach the five-mile mark before the end of the game with Lucky Luke ahead in points and Black Rain in the lead.

Black Rain starts to showboat showing off his flying skills turning his Drone side to side as it gets near the end. Black Rain continues to ignore the sensors of his Drone. Lucky Luke increases his altitude with another flurry of combos thrusting his Drone at unseen speed gaining ground on Black Rain. Seconds before crossing the finish line, Black Rain's Drone brightly light sensors all come on at once, **(*extreme damage has been done to left wing shutting down system now in three...two...one)*.** Black Rain crosses the finish line before his Drone completely shuts down. Unknowingly to Black Rain, Lucky Luke crosses the finish line at the same time.

The judges all gather to review the video. The panel of judges takes time to watch the video footage while the audience debates amongst each other. They conclude that the video shows Black Rain ahead of Lucky Luke until his drone shuts down and Lucky Luke is the winner by the nose of his Drone.

The crowd erupts with chants of LUKE ... LUKE ... LUKE!!

Black Rain removes his headset in revulsion walking furiously toward Lucky Luke. Luke removes his headset and stands in front of Black Rain. Lucky Luke crosses his arms under his chest and stares into the audience.

Director Mills panics, "OMG, What is happening here? We have millions of viewers this can't happen," with a look of disdain and concern.

Black Rain extends his hand to Lucky Luke, "Good job Rookie."

"Thanks! You did some outstanding maneuvering back there." Luke replied.

Sho-Gun is furious that Black Rain lost to Lucky Luke.

Mr. Alex stunned, "There you have it, Lucky Luke will advance to the championship round after defeating Black Rain. Everyone enjoy the down time get something to eat, drink or try the drone stations. One hour from now we'll see who will be in the championship round with Lucky Luke. Will it be nice sweet Angel or Sho-Gun?"

Director Mills turns walks over to Black Rain and Sho-Gun.

"What the hell just happened?"

"Don't even trip … I'll win the next round. Then I'll beat the brakes off Lucky Luke." Sho-Gun replies.

"That kid got some *hellava* moves, and he knew combos that I have never seen before." Black Rain replies.

"You sure as hell better. Just wait until the Admiral hears about this. OH MY GOD, he's going to be pissed." Mills walks away.

The media steps out to interview Lucky Luke.

"Hi, I'm Larita Duncan reporting live from the Drone's competition here at McCormick Place®. There was a major upset today in the

Drone competition. The favorite in the competition, Black Rain, was defeated by the local Chicago kid from the east side Lucky Luke as they call him. After today's events and winning by a very slim margin, I see why they call him Lucky Luke. I'm Larita Duncan reporting live for W.E.R.K. TV."

She signs off, "Back to you Dan in our studio."

In the studio, Dan Smith sits next to his female co-host Maria Santiago.

Dan responds to Larita, "You seem to be having a great time at the competition?"

Larita puts one hand over her ear to hear over the crowd, "Yes ... this is a lot more than I imagined it would be. The vibe here is *turnt* up. I never thought they would get a turnout like this one. I may even try the Drone experience."

Dan chuckles, "As you should ... as you should, a number of positive things have taken place around the city today. I'll tell you more after a word from our sponsor."

Sponsor, "Are you wanting that all-day fresh breath? Pick up a pack of Just Mint 4 U. Just Mint 4 U gives you a chance to offer someone a mint without offending them. Have one Just Mint 4 U. The polite way to recommend a mint."

Over in Admiral Greene's office, he is sitting at his desk watching the news when the phone rings. He answers, "Admiral Greene."

"Hello Admiral Greene. I just called to inform you that Black Rain lost."

Admiral Greene grits his teeth, "I know. I saw it on the news. I'm going to need you in my office for a full debriefing ASAP. Bring Ben with you."

"Who?" Director Mills asked.

"Black Rain as you call him." Admiral Greene replied.

"Yes, Sir."

Admiral Greene ends the call then tosses his phone into the wall. Immediately his secretary runs into his office.

"Is everything okay Admiral?" She asked.

Admiral Greene does not say a word he stares down at his broken phone. She turns quickly leaving out of the office.

At W.E.R.K. studios airing live Dan Smith and co-host Maria Santiago closing out their segment of the news as Larita Duncan watches.

"Closing out our news for the evening, the Chicago Bulls have acquired a power forward from overseas. He is supposed to make a big impact to help the Bull's get back to winning." Dan announced.

Maria Santiago with a huge smile, "That would be huge for the city of Chicago."

They sign off the air.

<u>One Day Earlier</u>

Larita walks up on the set where Dan and Maria Santiago are.

"Great Job team, you guys are good," Larita stated.

They all stroll back to the break-room area chatting on the way then takes a seat. Dan gets up rather quickly, "I have to get me a cup of coffee. Anyone else care for a cup?"

Dan returns with his cup of coffee. Larita smirks, "I am coffee, dark and hot"—Dan sips his coffee—"you sure are honey."

Maria laughs, "Enough already."

Dan takes another sip of his coffee, "Do either one of you find it odd that no one has taken responsibility for those bombings overseas?"

Maria with a serious expression, "You know what ... you're right about that. However, if

I bombed someone, I wouldn't say anything either."

"I second that. Dan maybe you should cut back on the coffee or drink decaf … you're getting a little paranoid," Larita stated.

"Yeah, Dan you're over analyzing the situation, trust and believe someone wants to take credit for it." Maria exclaimed.

"I just find it oddly strange that nothing has come out. I bet that you didn't know that those bombings have been going on over a month. A couple of them happened during the timing of the Drone contest."

Maria says, "Dan, I know you're not serious. Don't do that … to those kids, they're having the time of their lives with that competition."

"Dan has officially lost his mind those kids wouldn't know the first thing about bombing East Africa," Larita stated.

Dan in a daze holding his cup up to his lips as Larita stands pushing her chair under the table.

Maria utters, "I'll see you tomorrow Larita. Dan, I guess I'll see you as well Mr. Conspiracy." Maria laughs.

Dan in deep thought. Larita yells, "DAN!!!"

Dan spilling his coffee, "OMG ... What is it Larita?"

Maria laughs, "She was talking to you and you were somewhere else."

Maria and Larita laugh hysterically.

"See you tomorrow Dan" Larita asserted.

"See you tomorrow worrisome." Dan said with a smile.

Maria stands to leave, and Dan still has a troublesome expression on his face.

"Dan, if you really feel that strongly about it, check into it."

Maria walks away as Dan puts his coffee down on the table, "Yeah, just maybe I will." Dan replied.

Moments later Dan is sitting in the editing room watching old footage from the attack in East Africa. He selects English caption button as he reviews. The leader from the tribe is filled with emotion as he speaks about his horrific experience. Dan watches intensely as the chaos ensues, caption reads,

IT QUIETLY APPEARED FROM NOWHERE. MY PEOPLE SAY WHATEVER IT WAS HAD BEEN THERE FOR A WHILE. WE HEARD A LIGHT WHISTLING SOUND BUT SEEN NOTHING. WE CONTINUED HARVESTING OUR

D.R.O.N.E.

<u>Death Rules Over National Enemies</u>

CROPS AND ALL OF A SUDDEN, THE SUNLIGHT TURNED DARK. BY THE TIME WE COULD REACT THE WHISTLING SOUND GOT LOUDER, AND BOMBS EXPLODED. WHO WOULD DO SOMETHING LIKE THIS TO MY TRIBE?

Dan signals for the Chief Editor Scott Wilder. Scott quickly walks over to Dan.

"How can I get the timeline on this attack?" Dan asked.

"Dan, I'm in the middle of something at the moment. You need it right now?"

"No, but when you get a moment, can you please pull that footage along with Larita's coverage from the DRONE competition?" Dan asked politely.

"Sure Dan, just what are you up to?"

Scott rolls over his chair next to Dan taking a seat. Scott quickly starts to navigate through the footage with ease.

Scott explains to Dan how to follow timelines. "First, you type Drones competition in the box then click on search. Now click on the timeline and hit search again. There you have it."

Dan carefully reviews footage from the Drone competition. He slows down the footage spotting Director Chris Mills at the competition.

"Hey Scott, could you come over here for one more time just for a second?"

"Dan, come on I have to get this editing done for tonight's show"—Scott rolls over in his chair—"what is it now?"

Dan points to the monitor, "Isn't that Director Mills?"

Scott leans closer to the screen, "It sure looks like him. Let me zoom in. Yes, that's him alright."

They both speak at the same time, "Now what would he be doing at a gamers competition." They both stare at one another.

Scott starts to roll away, "Hey it's not illegal for him to be there. However, if you can find out why he's there that is the million-dollar question."

Dan abruptly jumps up exiting the editor's room without a single word.

Scott yells, "Dan what's the hurry? You left your coffee."

The next day over at McCormick Place® the Drone competition is underway. Sho-Gun is in the lead by more than thirty seconds over Angel. Sho-Gun seems to have this in the bag when suddenly Angel makes her move cutting the time down to fifteen seconds. Angel strikes a five-button combo gaining on Sho-Gun getting under ten seconds behind. Sho-Gun strikes a seven-button combo sending him into hyper-speed.

The crowd is silent having never seen hyper-speed before as Sho-Gun crosses the finish line twenty seconds before Angel. The crowd is silent for a moment with mixed emotions about Angel's loss. Angel removes her headgear. Mr. Johns walks over to raise Sho-Gun's arm.

The crowd upset starts booing, "BOO!!!!!"

Someone shouts from the audience, "Sho-Gun shouldn't have been allowed to play. BOO!!!"

Mr. Johns steps to the edge of the stage raising Angel's arm.

The crowd erupts with cheers, "Angel ... Angel ... Angel."

"Alright, all of you made your point. Angel is no loser. She will still receive a check for making it to the semi-finals. However, this only happened because of the social media presence surrounding the addition of Sho-Gun and Black Rain. My team and I heard you loud and clear."

Angel excited waves at the crowd and the crowd responds with chants of, "Angel ... Angel ... Angel."

"Now you have it the championship round Lucky Luke takes on Sho-Gun the assassin for the grand prize," Mr. Johns announced.

Angel walks over stands beside Lucky Luke and Sho-Gun is joined by Black Rain.

Later that evening over at THHS (The Hip Hop Spot) the official spot where all gamers hangout after the competition. The place is packed with a great vibe. One of the hottest local Rap Stars is in the building with his crew. Music fills the room as the crowd watches with enjoyment. Lucky Luke and Angel are close to the stage. She glances around the room checking the vibe when her eye's lock on Sho-Gun. Angel frowns as Sho-Gun walks in their direction.

"If it isn't loser and the next big loser sitting together," Sho-Gun stated.

Black Rain walks up behind Sho-Gun, "He got away from me Bro, but you got his number right Sho?"

"For sure that," Sho-Gun replied.

Angel angry, "I'd like to slap that stupid little grin off your ugly face, but the ugly is stuck there."

"Woo ... tough words for someone so small. I was almost scared," Black Rain stated.

Lucky Luke just sits there not paying any attention to Sho-Gun and Black Rain. Ace and a few buddies join in on the action. "Is there a problem over here Angel?"

Angel responds, "No, not at all," rolling her eyes in Sho-Gun's direction.

Sho-Gun smirks, "I came over to congratulate Lucky Luke on his win."

Black Rain burst out laughing, "Yeah they look like the new age Little Rascals."

Sho-Gun fist bumps with Black Rain, "Yeah Boy!!!"

Ace had enough of Black Rain and Sho-Gun stepping up into Luke's face.

Ace anger gets the best of him. "How about it?"

Black Rain laughs, "How about what?"

Sho-Gun steps in front of Black Rain. Black Rain steps from the shadow of Sho-Gun. Sho-Gun and Black Rain laugh at them.

Ace grits his teeth.

Angel upset, "Luke, you don't have to let them bully you."

Luke slightly embarrassed, "You don't have to defend me. I got this."

Sho-Gun giggles, "Look Black, Little Luke panties are showing."

Luke stands up very quickly drawing the attention of security.

Sho-Gun sees security coming grabs Black Rain's shoulders, "Let's roll."

"Yeah, we'll be seeing you and your man dressing girlfriend around." Black Rain stated.

Luke without hesitation, "You a Goofy... we'll see tomorrow who's wearing Vicki Secret."

Ace and the guys start to laugh as security walks up.

"I don't know what's going on over here, but save the drama for your mama. Take it to her house." Security said.

Black Rain walks off. Sho-Gun whispers to Ace, "I'll be seeing you around."

Ace frowns in the direction of Sho-Gun, "I wouldn't have it any other way. Don't look too far."

Everyone surprised by Lucky Luke's reaction. Angel looks at him as if he was her hero.

"You know that I can defend myself, right?"

"Yeah, I know. I'm really just tired of those bully type personalities." Angel stated.

"I may be a person of a few words, but I'm full of action." Luke stated.

Ace smiles, "Way to stand your ground Bro." Ace and his friends walk away talking about Black Rain and Sho-Gun.

Black Rain and Sho-Gun standing outside of THHS when Black Rain's phone rings.

"Hey! This is Director Mills. I'm going to need you to come over to Admiral Greene's office ASAP! And bring Sho-Gun with you."

"Yes Sir!" Black Rain replied.

About a half-hour later, Mr. Johns, and Director Mills are meeting with Admiral Greene when Lieutenant Belmer escorts Black Rain and Sho-Gun into the meeting. They all stand and salute one another before taking a seat.

"I'm not really sure if you gentlemen know why you're here? Make no mistake about it, it's a matter of National Security. Over the past few days certain things have transpired. Anything said in this room shall not be repeated under any circumstances. Benji Smith a/k/a Black Rain, you were recruited for this Drones competition based upon your flight time at the academy. Henry Lee a/k/a Sho-Gun, you were recruited because of your gaming ability, for whatever reason Black Rain lost. It was thought that you two would meet in the finals. Nevertheless, that won't be happening." Admiral Greene stated.

Black Rain and Sho-Gun glance at one another completely baffled.

"Admiral Greene, may I interject for a minute."

"Yes, go ahead Director Mills."

"I had chosen both of you guys in this competition to be winners by any means necessary. This operation is of great importance. I repeat by any means necessary. Sho-Gun, you have to win this competition. We have armed your drone with a new weapon unknown to your opponent." Director Mills exclaimed.

"Umm ... I'm a little confused here, we are talking about a video game?" Sho-Gun asked.

"I'm totally in the dark over here as well." Black Rain stated.

Director Mills leans forward in his seat, "Let me elaborate ... your careers are seriously at stake here, so I advise you to listen closely, because losing isn't an option. Winning is our only objective. The loss that Black Rain suffered compromised our mission."

D.R.O.N.E.
Death Rules Over National Enemies

Mr. Johns leans forward passing some documents to Sho-Gun. Sho-Gun scans through the documents quickly.

"To ensure that you win Mr. Johns has some information that will give you a competitive edge." Admiral Greene added.

"These are combinations to the Drone. But why?" Sho-Gun asked.

"You are correct. These combination codes haven't been released to the public; however, those Drones are more equipped than anyone knows. You would have to play over fifty hours just to get those codes. I need you to pay close attention to the swarm combination. It will multiply points once released." Mr. Johns stated.

Admiral Greene very stern, "We have our mission. Henry Lee, are we clear on this mission, correct?"

"Yes Admiral. I can beat this kid without any assistance." Sho-Gun replied.

Admiral Greene sternly states, "Never underestimate your opponent or you have already defeated yourself. Have a good evening, gentlemen."

Sho-Gun and Black Rain salute and exits the room quickly.

Admiral Greene falls back in his chair, "Are you sure Henry Lee can beat Lucky Luke?"

"I'm one hundred percent sure with those codes he can beat anyone." Mr. Johns stated.

"Tomorrow is an important day for you and Klaevion Electronics. Operation Swarm shall prevail."

They shake hands and exit the room.

Lucky Luke and Angel are leaving THHS when a W.E.R.K. news truck pulls alongside of them. The side door suddenly slides open.

"Come on inside I have something to show you about the Drone's competition." Dan Smith extends his hand for them to jump in.

Lucky Luke baffled by Dan Smith, "What is this about?"

Angel jumps inside, "Come on Luke."

Luke reluctant to join them jumps inside the van. The door on the van closes quickly speeding down the street with Larita Duncan driving.

"Hey there young people," Larita said with a big smile.

After driving a couple blocks away Larita finds a parking spot and joins them in the back of the van.

"Okay now, what is this all about?" Luke asked.

Angel is fascinated by the monitors and gadgets in the van, "This is really cool."

Luke getting on edge "Tell us why we're here or take us back right freaking now!"

Dan quickly pulls up some footage, "Pay attention to this footage."

Lucky Luke and Angel watch ever so closely, Luke is unsure about why he is watching these terrible attacks in Africa.

He pauses for a moment staring at Larita and Dan.

Lucky Luke confused, "And what does this have to do with me again?"

Dan looks deeply disturbed, "What if I told you that the Drone competition was related to these attacks."

Lucky Luke nods his head in disbelief, "Naw ... it's just a game."

"I've been closely monitoring this war game, and figured out a way to test my theory." Dan Smith exclaimed.

"You got my attention"—turning his head toward Dan—"how can we test your conspiracy theory?" Luke snickers.

Larita looks at Angel then Angel looks in Luke's direction, "Just consider it for a moment, Luke." Angel said.

"It would all make sense why they brought the military to play a game that was meant for entertainment." Larita stated.

Angel upset says angrily, "Those dirty Mother Lovers! "

"Still doesn't mean that the attacks are real. You guys are losing it." Lucky Luke replied.

"Angel is there something wrong?" Larita Duncan asked.

Angel pondering on everything she just heard, "Luke, there could be some truth to this story. Keep an opened mind and hear Mr. Smith."

Lucky Luke cuts off her words, "Not you too OMG? Okay just for the sake of knowing, what is it you want me to do?"

Dan sips his coffee, "How many weapons are there on a Drone?"

"The more you advance by levels the more weapons you receive. I'm sure that many people haven't made it to the level that I'm on." Lucky Luke answers.

"That's great, because if I'm right you're going to need more than the basic weaponry. This is what you may need during the flight...." Dan stated.

Everyone is silent for a moment.

Dan continues, "Like I was saying these attacks have been staged. I've been studying this footage for the last seventy-two hours."

"We have to stop this treacherous behavior, if these accusations are true." Larita responded.

"Either way it's worth a shot." Angel said looking at Luke. Her eyes start to tear up.

"What's wrong Angel?" Luke asked.

Tears stream down her face, "Just think how many people I have hurt if this is true"—Angel's emotions takeover—"they're just innocent people," while talking through her tears.

Larita leans over to embrace Angel, "You didn't know Sweetie, shame on our Government for taking advantage of gamers to push their agenda 5G."

Dan sips his coffee with a look of deception, "I did a little research on digital communication and computers. What I found is that all digital communication signals can be disrupted by an EMP."

"Okay, somebody help me out here. What's an EMP." Larita asked.

"It's like Superman's kryptonite for digital communication." Luke answered.

"I'm impressed that you knew that Luke." Dan replied.

Angel twist her neck and pokes out her lips, "Electric Magnetic Pulse."

Larita smiles, "Okay girl power."

Lucky Luke freaks out, "Wait a minute … no one in here knows for sure that it's true. Before we start throwing shade at the Government let's find out. This is lame."

"All the evidence would suggest that it's true." Larita replied.

THE CHAMPIONSHIP ROUND

The next day at the Drone competition it is a packed house with gamers from all over the world. The online attendance is up by thirty percent. Never has a video game brought together so many different age groups and nationalities. The media presence is tremendous for this championship round.

Mr. Roosevelt Johns walks on the platform and looks out into the sea of people gaming around the room. Mr. Alex is close on his heels.

"This had better work." Director Mills whispered to Mr. Johns.

"Relax … we got this one in the bag." Mr. Johns replied.

"May I have your attention please … may I have your attention just for a moment?" Director Mills asked.

The music halts and the games are all frozen. The final match takes place inside of a boxing ring. The spotlights are flashing inside the ring.

"The moment that all of you have been waiting for is near. What's at stake the largest gaming prize in gaming history? Never has a game had such success as DRONE. On behalf of myself and Klaevion Electronics CEO, Mr. Roosevelt Johns, we'd like to thank all of you for making DRONE number one in the world not just the country."

Mr. Johns steps up and grabs the mic with the upmost confidence, "GAMERS! I SAID GAMERS ARE YOU READY? Here in my hand are the fourth place and third place checks. Before we can give out any checks, we must see who our first-place winner will be. I SAID ARE YOU READY? Coming into the ring at five feet two inches and one hundred thirty pounds none other than sweet sensation Angel Chung a/k/a The Angel."

The crowd erupts in cheers and chants of "ANGEL ... ANGEL ... ANGEL."

Angel makes her way up on the stage walking like a model. She is wearing stretch jeans and fitted T-shirt with her hair done and wearing light makeup.

"I see you girl power." Larita said from the front of the stage.

Luke admires Angel's beauty from behind the curtain, "I see you girl ... DANG!"

"Coming in from the opposite side of the ring at five feet nine inches and one hundred seventy pounds is none other than Benji Smith a/k/a Black Rain." Mr. Johns announced.

The crowd boo's "BOO ... BOO ... BOO...."

"Black Rain is a fraud." Someone yelled out.

Angel fans her hands for the crowd to settle down. The crowd follows Angel's lead. The media is filming, and cameras are just flashing as Black Rain and Angel stand side by side.

"Before the day is over you will know who has first place and second place. Black Rain and The Angel have already secured third and fourth place." Mr. Johns stated.

Dan Smith is standing in the back of the arena looking on. He is slightly irritated by a fly who keeps trying to land on him. He keeps swatting at the fly as he ducks away. Dan is focusing on what's about to take place. The fly lands on Dan's back without him knowing.

"OUCH!" Dan said while popping the back of his shirt. He reaches under his shirt to rub his finger across the bite.

Meanwhile, Lucky Luke and Sho-Gun enter the ring. The lights in the ring are flashing. Lucky Luke is rocking skinny jeans and Timberland® boots with a matching Timberland® T-shirt. Sho-Gun is rocking a Jordan® sweat suit and late model Jordan® tennis shoes. Mr. Johns stands between Lucky Luke and Sho-Gun as the crowd cheers.

Mr. Alex tries to get the crowd more pumped as he yells, "IT'S ABOUT TO GO DOWN!"

"Yes Sir," Mr. Johns responded.

"Today's sponsor BunS has provided us with two beautiful ring girls. The BunS model walks out wearing stretch denim shorts and tank top with the BunS logo on the front. She's carrying a tray with virtual reality gear ("VRG"). She stops in front of Lucky Luke, he takes one. She turns toward Sho-Gun and he grabs the other.

"Isn't she a caramel beauty? Thanks Ms. BunS, calendar girl of the month and runner up."

They smile and bow their heads exiting the ring.

"Both of you guys have your virtual reality gear. This final round will be done with that gear … no more looking at the monitor and the crowd. It's tunnel vision from this point on. The only thing you can see is what's in front of you." Mr. Johns said with a slight giggle.

"Gentleman, take a seat center stage and place your virtual reality gear on." Mr. Johns stated.

Sho-Gun sits first grabbing his controller looking it over. Luke sits next grabbing his controller and glances out into the audience.

Luke's face appears to be made out of stone until his eyes cross Angel's. He smirks and winks at her.

"Hello Gamers! I'm Larita Duncan reporting live from the Drone's competition. Never in all my years of working in media have I seen an event like this Drone competition. This competition has been spectacular from beginning to end. With that said, the end is near for one of these contestants."

Mr. Johns speaks with great excitement, "Place your VRG on and listen for the countdown."

Lucky Luke glances around the ring a second time to see Angel with her eyes closed and hands folded in front of her. Sho-Gun glances around the ring to see his guy Black Rain.

Black Rain is nodding his head up and down. Sho-Gun gives Black Rain a thumbs up and slight nod.

Mr. Johns says, "Global Gamers around the world sit tight as Drones take flight. Start the countdown."

The crowd goes wild with chants of Luke ... Luke ... Luke ... Luke! Gun ... Gun ... Gun ... Gun ... Gun!

Lucky Luke makes a final adjustment putting on his headphones and pulling his hood over his VRG.

"This final round will be a thirty-minute flight to gather up all the points you can. The more you kill the more points you gain. The first one to the end of the level is the winner unless the other person has the most points; in that case we will have a draw and have to tally point's verse's time."

"Is the timer ready?" Mr. Alex asked.

Mr. Johns glances overhead at the scoreboard with the timer. "The timer is set."

"Ten…nine…eight…seven…six… the crowd joins in the countdown 'five … four … three … two … one!'"

The crowd erupts as the Drones take flight with the sound of a light BUZZ!!

Sho-Gun takes off quickly in super trust mode leaving Lucky Luke behind. Both Drones enter the kill zone and Lucky Luke opens fire missing everything. Sho-Gun lets off a Smart-bomb blowing up everything in sight, he scores big with a direct hit on a fishing boat. Lucky Luke enters over the Red Sea gaining ground on Sho-Gun. Sho-Gun continues to rack up points as Lucky Luke is trying to gain ground.

The audience is stunned by Lucky Luke's behavior, everyone around looks at one another trying to figure out why Lucky Luke has not opened fire? Lucky Luke adjusts his VRG and spots Larita Duncan and Dan Smith. He wipes a few beads of sweat from his forehead.

Director Mills receives a call from Admiral Greene.

"Hello!"

"Mills, I just received some intel from one of my field officers that Luke has been meeting with that reporter Dan Smith."

"What about Admiral?"

"It doesn't matter to me; we can't have any loose ends."

"Yes Sir, Admiral."

"I've already released one of the Swarms mini as soon as I got the information. It should take affect within the next five or ten minutes."

As they speak Dan Smith collapses on the side of the ring.

"OMG! Help!" Angel screams!

Dan Smith is unconscious surrounded by gamers as they look on to see what just took place. The commotion was so loud that Sho-Gun and Lucky Luke both peaked from under their VRGs. Larita Duncan in a state of panic locks eyes with Luke. Lucky Luke nods in her direction. He grits his teeth together and releases a flurry of combos hyper-spacing him within seconds of Sho-Gun. Sho-Gun continues killing everything in sight gaining points as he maintains his lead.

Director Mills concerned, "Admiral, I've noticed quite a bit of eye contact between Luke and Larita Duncan."

"No worries, we'll nip it in the bud. Dan Smith is down … that's all that matters."

Lucky Luke and Sho-Gun enter over farmland and Sho-Gun opens up a flurry of combos with gunfire and Smart-bombs. Lucky Luke right on Sho-Gun's tail smirks under his VRG.

The paramedics haul off Dan Smith with Larita by his side. Larita walks along side of the gurney holding Dan's hand. Dan's hand falls limp slipping out of Larita's hand. She's filled with emotions, "OH MY GOD! NOOO!!"

Angel instantly tears up thinking about what Dan shared with them. Angel rushes over to Larita and embraces her.

Director Mills returns to the ring abruptly trying to gain control of the atmosphere. "Gamers there will be a brief intermission until we can get things under control. The contest has been paused. Lucky Luke and Sho-Gun will resume shortly. Thank you for your cooperation. Please be as patient as possible while the paramedics walk through."

Larita Duncan receives a call from W.E.R.K. "Yes, this is Larita."

"We need to go live from the competition ASAP. The competition is trending on social media with people streaming on FACEBOOK, TWITTER, SNAP, Tik Tok, and IG for a live feed."

Larita replies, "I understand, but Dan ... wasn't looking to well."

"Let us worry about Dan. We need to air this story."

Angel gives Larita some napkins to wipe away her tears. Larita tries to regain her composure taking a deep breath. The camera man is set to air.

Angel hugs Larita, "You got this."

Larita takes a second-deep breath and points to the camera man. He begins his countdown to air in three ... two ... one.

Her voice is shaky, "Hi! I'm Larita Duncan reporting Live from the DRONE finals here at McCormick Place®. Today has been a day filled with excitement. The competition has been placed on hold due to the nature of one of my co-workers taking ill during the competition. 'Dan

Smith, we are all praying for you and your loved ones.'"

Larita starts to tear up, "I'm Larita Duncan reporting live from the DRONE competition. Now a word from our sponsor."

A Sponsor commercial appears on the screens, "JMFU, Just Mint 4 U, is a refreshing chewing sensation that freshens breath like no other leaving your mouth feeling clean."

Lucky Luke walks over placing his hand on Larita's shoulder. They walk through the crowd as people patted Luke on the back.

On the other side of the room Black Rain and Sho-Gun are talking about the competition.

Black Rain unfazed by the moment, "What the hell is going on? Some reporter Dude takes ill and they pause the game?"

Sho-Gun giggles, "Yeah, where they do that at? Don't worry I got this though."

Black Rain laughs, "I know Bro! How long of a break did they say you would get?"

"They didn't, but I'm ready whenever." Sho-Gun replied.

Black Rain's attitude changes slightly, "Bro, there's been some rumors floating around the competition."

"Like what?" Sho-Gun asked.

Announcement on the PA, "GAMERS ... WE WILL RESUME IN FIVE MINUTES. FINALISTS TAKE YOUR POSITIONS IN THE CENTER OF THE RING. THE COMPETITION MUST GO ON."

Sho-Gun fist bumps Black Rain, "I'm out Rain."

"Yeah, handle that Dude." Black Rain shouted.

"What was the rumor?" Sho-Gun asked.

"It was nothing." Black Rain stated.

Back in the middle of the ring the game resumes.

Sho-Gun picks right off where he left off blasting everything in sight. Lucky Luke seems to be more concerned with catching up to Sho-Gun then racking up points. The Drones leave from over the sea and enter over farmland.

It is a warm fall day in East Africa. Another small tribe in Somalia is out harvesting their crops when a sudden shade appears from above. The workers stop working and take note of the sudden shift of shade. This time one of the tribe members starts running and the others follow including the tribal leader swiftly scattering in all directions quickly taking cover from this large unidentified flying object.

Sho-Gun's Drone goes to autopilot without his assistance and he reads out loud, "Activating Operation Swarm, is flashing on my dashboard."

Sho-Gun tries to control his Drone without any success, "What the heck is going on here?"

Operation Swarm in progress.

"WHAT THE HECK!" Sho-Gun struggles to regain control of his Drone. Flying over farmland with villagers, Sho-Gun's rear deck opens dropping out something that appears to be like Bee's. The Swarm of Bees descended toward the farmers. Lucky Luke looks on through his VRG closing in on Sho-Gun. Lucky Luke lets off a seven-button combo while flipping his controller in a three hundred sixty degrees motion. The maneuver stuns the audience because no one has ever attempted to attack another Drone. Lucky Luke approaches quickly from the rear with a flurry of missiles that slows Sho-Gun for a moment. The Swarm of Bees gets close to the villagers. Lucky Luke makes his descent down behind the Bees. Lucky Luke releases a magnetic pulse just as the Bees start attacking the villagers. The villagers let out loud screams as the Bees fall to the ground after landing on them. Sho-Gun uncertain about the auto-pilot resumes control of the Drone tries to elude Lucky Luke.

Director Mills and Mr. Johns look at one another with confusion.

Mr. Johns in disbelief, "How is this possible?"

"I don't know but you had better figure it out quick." Director Mills exclaimed.

Sho-Gun is on a mission trying to distance himself from Lucky Luke's Drone.

Lucky Luke reaches down turns his music up in his headphones and lets off an arsenal striking Sho-Guns Drone. Sho-Gun drops his controller as it vibrates. His Drone goes down in smoke. The audience erupts in cheer when Sho-Gun's Drone crashes to the ground. Lucky Luke races for the finish-line. The moment Lucky Luke crosses the finish line the crowd erupts with chants of LUKE ... LUKE ... Luke. Lucky Luke removes his headphones and his VRG. Angel rushes up to the stage giving Luke a big hug and kiss him in the mouth. Lucky Luke raises his controller over his head.

Lucky Luke yells, "YES!!! I'm the Champion!"

Sho-Gun with a look of disgust drops his head.

Black Rain goes to Sho-Gun, "Bro, pick your head up. It's just a game."

Sho-Gun looks up at Black Rain without speaking. Lucky Luke looks over at Black Rain and Sho-Gun with a look of distain.

"Collusion or delusion anyone can get it anytime ... my house!" Lucky Luke stated.

Director Mills stares at Lucky Luke wondering how much he really knows. Mr. John's cellphone rings. It is Admiral Greene.

Admiral Greene sits in his office rocking back and forth.

He yells, "JOHNS ... SOMEONE PLEASE TELL ME WHAT JUST HAPPENED?"

"Admiral Greene, we had some unforeseen situation occur." Mr. Johns answered.

"WHAT THE HECK DOES THAT EVEN MEAN? ALL I KNOW IS THAT WE HAVE A MILLION DOLLAR AIRCRAFT SOMEWHERE OVER IN AFRICA AND YOU BETTER WORK LIKE HECK TO GET IT BACK. AM I CLEAR"—Admiral Greene pounds his fist on the table several times as he speaks—'DEATH RULES OVER NATIONAL ENEMIES.' ANYONE STANDING IN THE WAY OF OUR FREEDOM IS DEAD TO ME. IF YOU AND DIRECTOR MILLS DON'T FIGURE OUT A WAY TO RETRIEVE THAT DRONE YOU HAD BETTER MOVE TO SOMALIA AND PIECE IT BACK TOGETHER!"

Lieutenant Belmer enters the room unannounced.

Admiral Greene frowns and yells, "WHAT IS IT NOW LIEUTENANT BELMER? CAN'T YOU SEE THAT I'M IN CRISIS MODE?"

"Excuse me Sir, the White House is on the line it's the President, Sir."

Admiral Greene looks as if he is about to implode as he looks around the room.

Back in Somalia the village is burning with some small casualties. The tribe leader stands next to the fallen Drone with his staff over his head. One of the small villagers walks over to the leader handing him one of the Bees. He looks it over, "It's metal!"

At that moment, the eyes light up on the Bee and his image is projected back to the Department of Defense.

THE END

Intellectual Properties of
KREW ENTERTAINMENT GROUP LLC

KREW Bangers™

Just Mint 4 U – JM4U™

TTHS - The Hip Hop Spot™

Winners Ice™

BunS™

**KEVIN WHITAKER
KREW ENTERTAINMENT GROUP**

PRESENTS

D.R.O.N.E. II
DEATH RULES OVER NATIONAL ENEMIES

Dan Smith is in recovery coming out of his coma like state. He finds Luke, Larita and Angel by his bedside.

Dr. Johnson stands there with her clipboard. "Well, Mr. Smith, I have good and bad news for you. The good news is there's no life altering illnesses here, but we haven't seen any cases like this in the United States. However, there have been cases reported similar in Africa. What troubles me is how you contracted the virus."

Larita, Luke, Angel and Dan all look at one another.

At Admiral Green's office, he is furious that Operation Swarm was not a success. He has all agencies in his office along with Mr. Johns of Klaevion Electronics.

The media has gotten word that the attacks on Somalia could somehow be related to the Drone's competition.

Admiral Greene pounds his hand on the table, "The question here is how do we fix this problem and recover our Drone? By whatever means necessary, GET MY DRONE BACK AND DO A MEDIA BLACKOUT!"

These titles are for mature readers.

CRACK'IN – Chris Miller a/k/a Brown, an intelligent man who started out making fake IDs as a teenager, has moved to the next level with his business. Chris has stolen credit and credit cards using alias names. He meets a young computer geek, Octavius, who has no social life. Chris introduces Octavius to a world filled with women, cars, and money. Octavius is fascinated by the lifestyle Chris has shown him. Octavius is clueless about what is transpiring after Chris convinces him to intercept a wire transfer from one of the mobs' accounts. The Grasso family of New York hires a hacker to track down their money, leading them to the City of Chicago.

The Party Girl – This awesome story will take you through peaks and valleys while keeping you ever so entertained. It is filled with everything a high school student may experience from humor, drama, deceit, passions, jealousy, and betrayal.

SCORN The Legacy – "Scorn," Eddie Fox and his beautiful wife Sheila's three boys are, Eddie Jr.,

Roosevelt, and Victor. The Fox family is living the American dream in the City of Chicago until tragedy strikes. Eddie Sr. has placed his family in harm's way when he takes on a debt, he cannot afford to one of the city's largest drug lords. Eddie Sr.'s decision making will take you on a roller coaster of action, lust, suspense, jealousy, deceit and betrayal. In today's family structure there is nothing like family love! "Scorn" is highly entertaining and most intriguing pushing the boundaries of family love to the limits.

www.ingramcontent.com/pod-product-compliance
Lightning Source LLC
Chambersburg PA
CBHW071538100726
47908CB00004B/1432